DATE DUE

JUL 2 6 2006		
OCT 0 5 2006		
NOV 2 8 2006		
DEC 2 9 2006		
FEB 1 0 2007		
APR 14 2007		
MAY 2 2 2007		
JUL 1 8 2007		
OCT 2 3 2007		
JAN 0 7 2008		
FEB 0 7 2008		
MAY 2 7 2008		
NOV 3 0 2008		
JAN 3 1 2009		
JUL 0 6 2010		
JUL 3 0 2010		
GAYLORD		PRINTED IN U.S.A.

Brave Norman

A True Story

First Aladdin Paperbacks edition November 2002
Text copyright © 2001 by Andrew Clements
Illustrations copyright © 2001 by Ellen Beier

ALADDIN PAPERBACKS
An imprint of Simon & Schuster Children's Publishing Division
1230 Avenue of the Americas
New York, NY 10020
Also available in a Simon & Schuster Books for Young Readers
hardcover edition.

READY-TO-READ is a registered trademark of Simon & Schuster, Inc.
The text of this book was set in Times New Roman.
The illustrations were rendered in watercolor.
Printed and bound in the United States of America
10 9 8 7 6 5

Library of Congress Cataloging-in-Publication Data
Clements, Andrew, 1949–
Brave Norman / written by Andrew Clements ; illustrated by Ellen Beier.
p.   cm.—(Pets to the Rescue)
Summary: Norman, a blind Labrador retriever, saves a girl from drowning
in the ocean.
ISBN: 0-689-82914-0
1. McDonald, Steve, 1957– —Juvenile literature. 2. Labrador retriever—
Anecdotes—Juvenile literature. [1. Dogs. 2. Blind. 3. Lifesaving.]
I. Beier Ellen, ill. II. Title. III. Series.
SF429.L3 C54 2000   636.752/7 21   99-039637
ISBN 0-689-83438-1 (Aladdin pbk.)

PETS TO THE RESCUE

Brave Norman

A True Story

Written by Andrew Clements
Illustrated by Ellen Beier

Ready-to-Read
Aladdin Paperbacks

New York   London   Toronto   Sydney   Singapore

Steve McDonald wanted
another pet for his family.
He went to an animal shelter.

4

He saw many dogs.
He saw many cats.

Then he saw Norman.

Steve said,
"That's the one!"

The shelter people said,
"No one wanted this dog.
We are very happy
you picked him."

6

So was Norman.

Norman loved his new family.

He loved Steve.
He loved Annette.

He loved little Paul.

And he loved Lucy,
the family's other dog.

Norman followed
Annette and Paul and Lucy
around all day.

He went to the park with them.
And he loved going
to the beach.
But one day Norman
started bumping into things.

The family took him
to the animal doctor.

The doctor said,
"This dog is going blind.
There is nothing we can do."

The family
took Norman home.

Friends said, "Poor dog.
Are you sure you want
to keep him?"

14

Annette said,
"We love Norman.
We will take care of him
no matter what."

It was not always easy
having a blind dog.

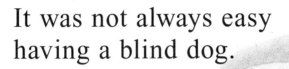

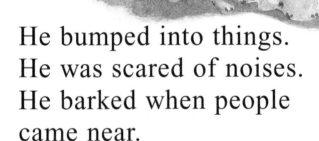

He bumped into things.
He was scared of noises.
He barked when people
came near.

Lucy was old and very kind.
She knew Norman
needed help.
Lucy walked close to him.
Then Norman did not
bump things.

Norman still went to the beach.

At the beach,
there were no trees to hit.
There were no cars
to scare him.

Norman ran and ran.

One day Annette and her mom
took little Paul to the beach.
Norman came too.

Suddenly Norman ran off.

He ran into the water
and started swimming.

Then Annette saw something.

A girl was far out in the water.
She was having trouble.
She had called for help.
No one had heard her
except Norman.

The girl kept calling,
and Norman kept swimming.

When Norman got close,
the girl grabbed his tail.

Annette watched Norman.
Norman was helping the girl.
But he was swimming
the wrong way.

Annette called, "Norman!
Here boy!"

Norman swam
toward the beach.
He was pulling the girl!

But the girl was very tired.
She let go of Norman's tail.

Norman tried to find the girl.
He swam around and around.

Annette yelled to the girl,
"Call him! His name is
Norman!"

"Norman! Norman!" called
the girl.
Norman came near.
The girl grabbed him again.
This time she did not let go.

Norman pulled the girl
out of the water.
People helped her
onto the beach.

The girl was cold and scared.
But she was safe.

Norman had saved her life.

Norman used to be the dog
that no one wanted.

People used to feel sorry
for Norman
because he was blind.
Now there are books
about him.
He has been on TV.

Being blind did not stop
Norman from being brave.
Being blind did not stop
Norman from helping.

Now Annette is training
Norman to visit hospitals.
Norman can make
children laugh.
Norman can help sick people
feel a little better.

Norman is a true animal hero.